MISSOURI PACIFIC LINES
6617
6617
MISSOURI PACIFIC LINES
FUR

Selected Photographs from the Missouri Pacific Historical Society Archives

by Charles A. Duckworth

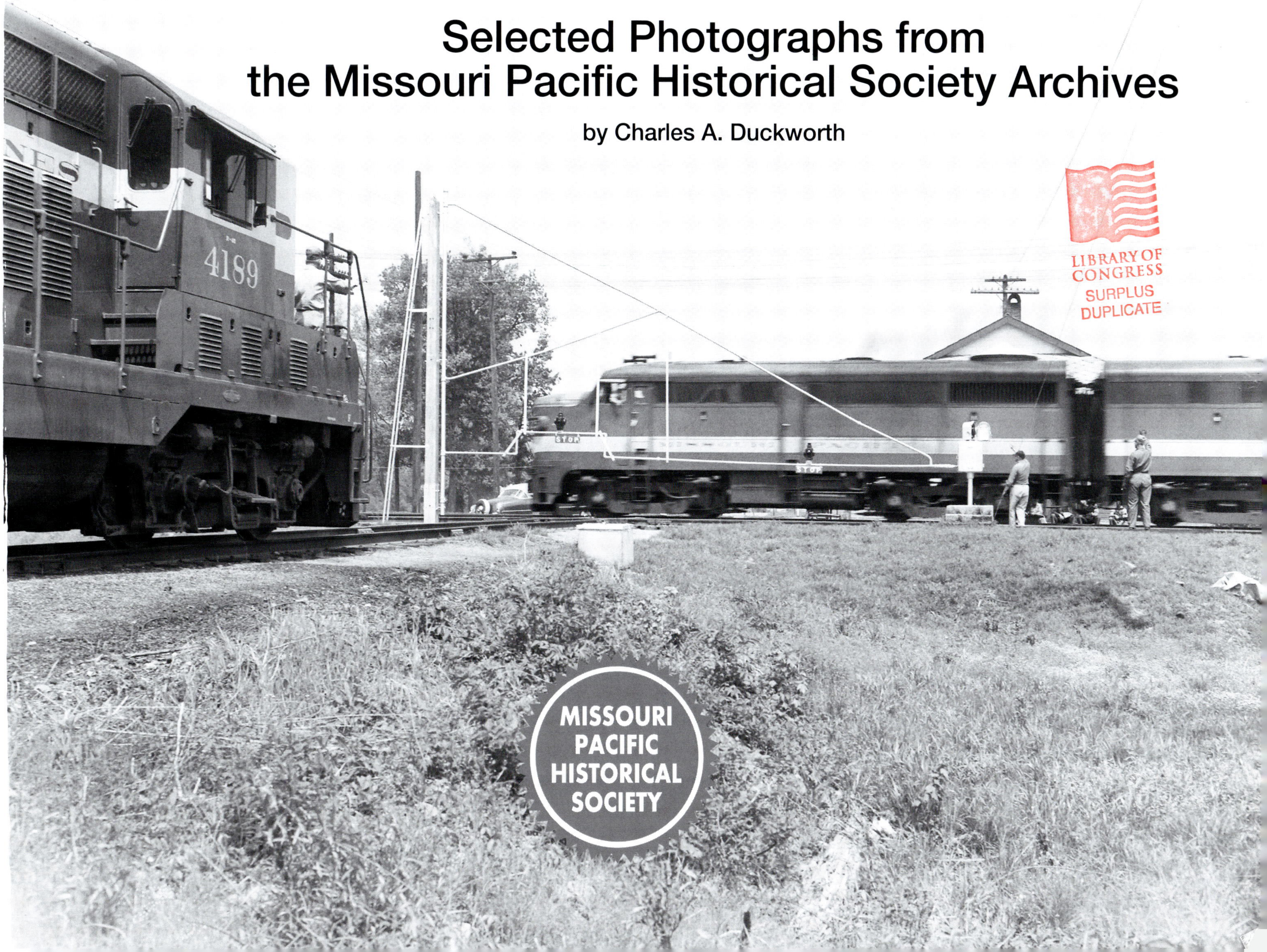

MISSOURI PACIFIC HISTORICAL SOCIETY

The Donning Company Publishers
731 South Brunswick Street
Brookfield, MO 64628

Lex Cavanah, General Manager
Nathan Stufflebean, Production Supervisor
Anne Burns, Editor
Chad H. Casey, Graphic Designer
Katie Gardner, Marketing and Project Coordinator

Steve Mull, Project Director

ISBN: 978-1-68184-318-6

Cataloging-in-Publication Data available from the Library of Congress

Printed in the United States of America at Walsworth

1701 on a thirty-one-car Dupo to St. Louis transfer at Valley Junction, Illinois, October 20, 1951. Joe Collias photograph

Table of Contents

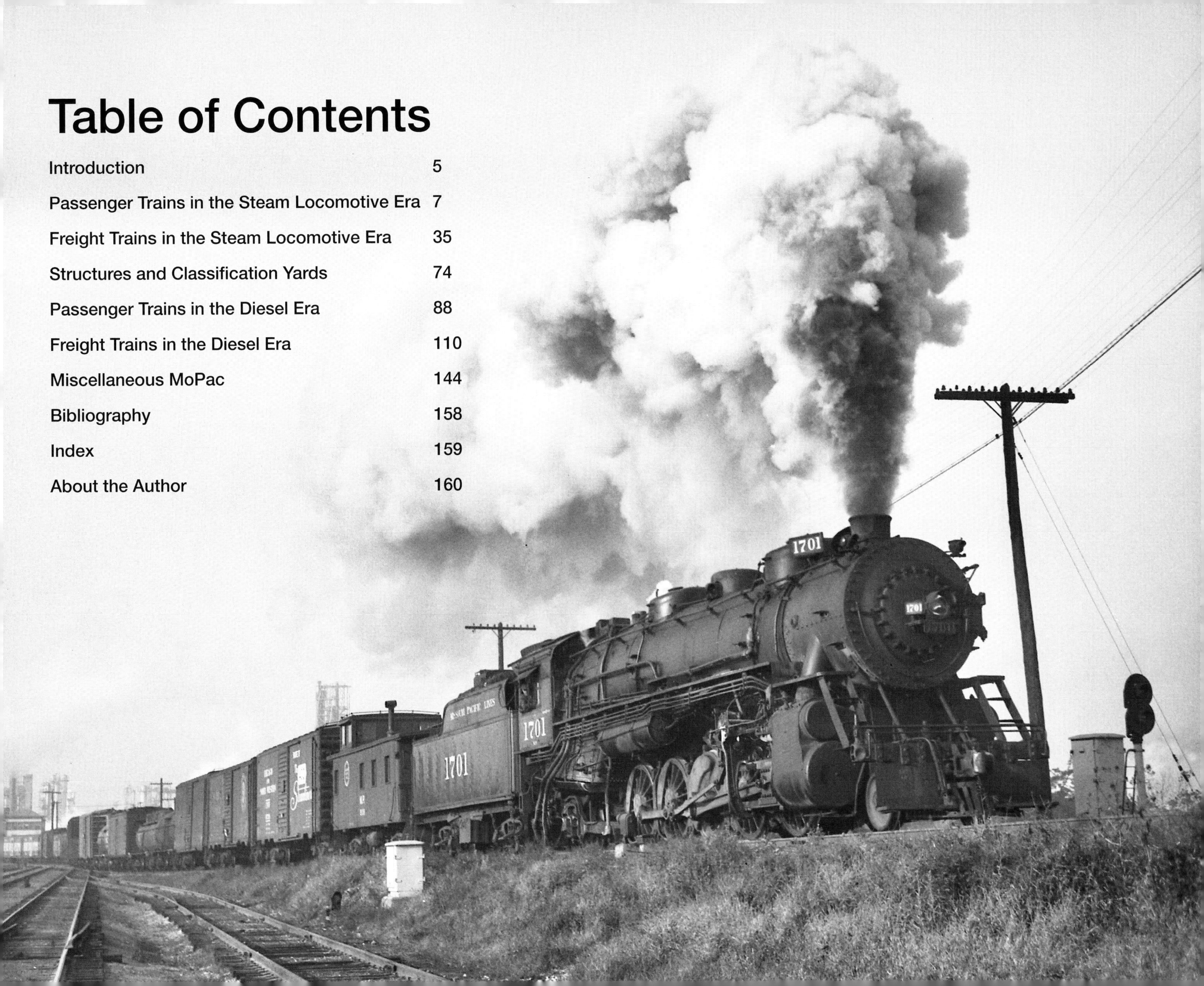

This book is dedicated to several of the deceased members of the Missouri Pacific Historical Society who were active supporters: Mike Adams, Joe Collias, Tony Fey, Bill Hanson, and Steve McVey.

1716 is helping 5344 train No. 79 westbound up Kirkwood Hill in January 1950. Joe Collias photograph

Introduction

In September 2018, the Missouri Pacific Historical Society purchased the Missouri Pacific portion of the Joe Collias Collection comprised of over five thousand negatives covering Missouri Pacific, Texas & Pacific, the Muskogee Roads, and Missouri Southern trains, locomotives, rolling stock, and structures. I'd first been introduced to his collection in 1975 when I met Joe for the first time when I was transferred to St. Louis by the MoPac. Joe was an avid rail photographer, historian, author, and award-winning scale modeler (having built detailed HO scale steam locomotives from brass and building complex resin freight car kits). I'd ring the doorbell, Joe or Marjorie would answer the door, and Joe and I would head for the basement for a couple hours where he'd sit at his work bench (an old oak roll-top desk) and I at his drafting table looking at his latest model creations or photograph prints. After a few questions on what my current "mission" was, Joe would pull out photographs or negatives to help with whatever model I was building or what article I was writing for a magazine. Joe was generous in making prints for this new enthusiast and obviously planted the seeds for my own articles and future books for the society. Living in Maplewood, Missouri, Joe began biking to the tracks as a teenager and took MoPac and Frisco images of the trains and locomotives with a box camera. After his military service, Joe traveled thousands of miles in the 1950s photographing the last railroads still running steam locomotives. These images became the foundation for his first two books, *Last of Steam* and *In Search of Steam*. He followed these books with *Mopac Power*, *Texas & Pacific Power*, *Frisco Power*, *Katy Power*, and *Mopac in Color*.

Joe was a traditionalist in the sense of photographing railroads. Following in the footsteps of Robert Foster, James Bowie, C. T. Wood, Lucius Beebe, and A. E. Brown, he preferred to capture his favorite railroads on film with either a locomotive roster image or a traditional three-quarter train "wedge" photograph with more smoke appearing all the better. Wayne Leeman, Joe's friend and fellow railroad enthusiast, gave Joe his negative collection when he retired. Wayne was a *Post-Dispatch* newspaper photographer for forty-six years. Wayne, being a railfan, used his newspaper credentials to ride and document train activity for photo features in the newspaper or in *Trains* magazine. Used in this book is his documenting Missouri Pacific's steam helper service over Kirkwood Hill, locals switching a mill in Sikeston, steam transfer runs between St. Louis and Dupo, the Illinois coal fields in 1955, and a trip down to Leeper, Missouri, riding with the head-end crew on an Alco passenger diesel. In addition to Wayne's images, Joe's collection also included negatives from A. E. Brown, C. T. Wood, R. S. Plummer, and others. After Joe's death in 2015, his wife Marjorie donated his photograph and negative collection to their church, Lutheran Church of the Resurrection, as a fundraiser. Recognizing that this was an opportunity to acquire one of the largest MoPac image collections in the country, David Huelsing, MPHS archivist, worked with the MPHS board members and other members to raise the funds to purchase and keep the collection intact.

In May 2021, I went through some two thousand images and chose a balance of photographs I felt the reader will enjoy. In composing the captions, I enlisted the help of several former Missouri Pacific employees and friends to provide additional details to my captions. Norbert Shacklette and Jerry Laseter provided background to the St. Louis and Sikeston photos, Tom Stamey helped with information on the Texas & Pacific images, Bill Pollard provided his expertise on the numerous passenger train images used in the book, and Dennis Faircloth provided his help with the Missouri–Illinois photographs.

This certainly isn't an in-depth history of the Missouri Pacific but hopefully an enjoyable photographic study of this large St. Louis-based railroad spanning eleven states and over eleven thousand miles of track. All images used are from the Missouri Pacific Historical Society/Joe Collias Collection with the captions crediting the photographers or known original sources.

Charlie Duckworth, Omaha, Nebraska

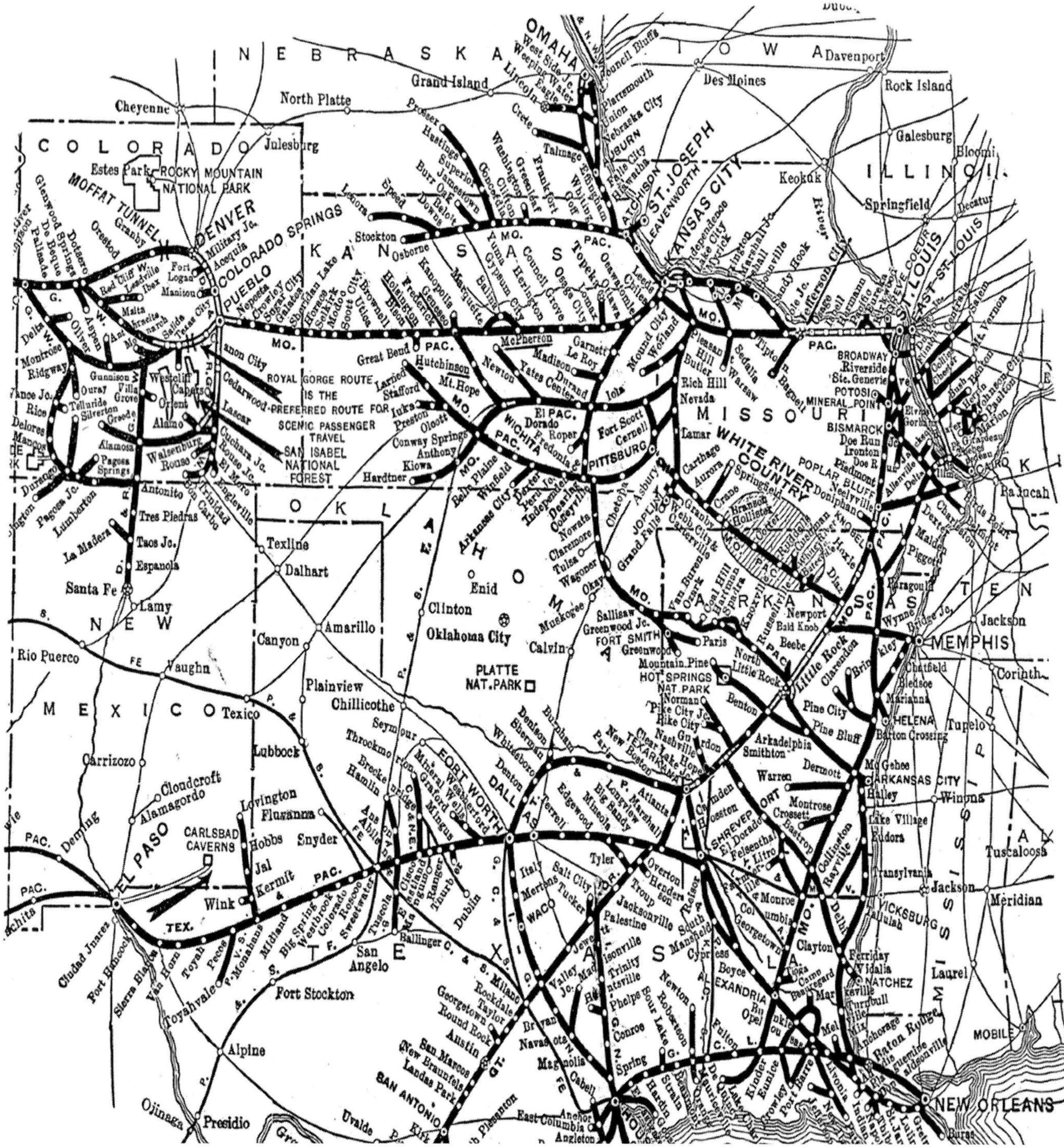

MISSOURI PACIFIC LINES

THE TEXAS AND PACIFIC RAILWAY
TEXARKANA · SHREVEPORT · EL PASO · NEW ORLEANS

This map from MPRR dates from the period in which the photographs in this book were taken. It includes the Missouri Pacific Railroad and the Texas & Pacific Railway.

Passenger Trains in the Steam Locomotive Era

6629 is on train No. 37 southbound near Fulton, Arkansas, on the Little Rock-Texarkana mainline in the summer of 1948. In earlier years, No. 37 had functioned as an overflow section of train No. 7, "The Southerner," carrying revenue passengers and several Pullman car lines. On January 4, 1948, however, No. 37 disappeared from public timetables but continued to operate as a mail and express train, appearing only in employee timetables. No. 37 was still considered a first-class train but carried only head-end traffic and a rider coach for the train crew and occasional deadheading employees. R. S. Plummer photograph

6611 is on train No. 3 southbound at Poplar Bluff, Missouri, in September 1951. MoPac's postwar efforts to repaint heavyweight passenger equipment into Eagle colors is apparent. Poplar Bluff was a major division point and the endpoint terminal for both the Arkansas Division and the Missouri Division. As such, the area around the depot was well populated with railroad hotels to accommodate train crews that were laying over between runs. Several of these hotels are visible in this view, including (left to right) the Crown Hotel, Dunn Hotel, Railroad YMCA, and Van Noy Hotel. For a period before the Poplar Bluff depot was constructed in 1910, the first floor of the Crown Hotel served as ticket office and waiting room for the Iron Mountain. Joe Collias photograph

In 1946, Missouri Pacific No. 7, "The Southerner," left St. Louis at 12:30 am and arrived in Little Rock at 10:05 am. At this time, the train carried a Chicago-Hot Springs Pullman off the Alton, as well as a St. Louis to El Paso Pullman which opened for occupancy at 9:30 pm. No. 7 is shown here with Mountain-type 5316 approaching North Little Rock's Main Street viaduct. Because of frequent congestion and convergence of many tracks into a small area, this part of the North Little Rock yards was known locally as "the bottleneck." On the north side of the tracks in the background can be seen parts of the large Little Rock Stock Yards. MPRR photograph

A nine-car mail Extra leaves Little Rock behind 6628 in 1942. The late afternoon lighting suggests this might be an extra section of train No. 8. The Sullivan-Renshaw postal car behind the tender, built with outside channels, was a unique design only used by Missouri Pacific and Texas & Pacific. MPRR photograph

6001 heads up southbound No. 17 with seventeen cars just crossing over the Rock Island mainline at HH Tower. Initially built as a three-cylinder locomotive, it was numbered 6000 in 1925 by Alco. Having three cylinders, maintenance was an issue and 6000 spent more time in the shops than in service, resulting in being stored from 1931 to 1937. Missouri Pacific rebuilt the engine as 6001 with poppet valves, roller bearings, and Elesco exhaust injector and converted it to an oil burner and two cylinders. The rebuild was a success with over 102,000 miles in operation the first twelve months. Photograph taken on May 5, 1951, by Johnnie Gray.

No. 231, the "Southern Scenic" has just left the Cricket, Arkansas, depot in this photograph taken in October 1946. At this time, the train carried an RPO–baggage, a coach, and a dining-parlor car between Memphis and Kansas City through scenic White River country. At Kansas City, through passengers made a cross-platform connection to trains 11 and 12, the "Colorado Eagle." E. G. Baker photograph

A. E. Brown caught the northbound No. 132, the “Louisiana Sunshine Special,” at Tioga, Louisiana, in August 1952. This section of the “Sunshine” originated at Lake Charles and made cross-platform connections with Houston/Galveston trains at Kinder, Louisiana. At this date, the “Louisiana Sunshine” carried a Lake Charles-St. Louis Pullman (eight section, five double bedroom configuration), Lake Charles to Little Rock coaches, an Alexandria to Little Rock grill coach, an RPO–baggage car, and assorted head-end traffic. At Little Rock, the St. Louis Pullman was added to train No. 32, the “Sunshine Special,” which now operated as a third section of the “Texas Eagle.”

Another view of No. 132, the "Louisiana Special," being pulled by Pacific-type 6614 at Alexandria, Louisiana, in November 1951. H. Stirton photographer

6607 on a mail and express Extra at California, Missouri, in August 1950. Missouri Pacific carried a variety of head-end cars off the Pennsylvania; New York Central; Chesapeake & Ohio; and Gulf, Mobile & Ohio Railroads and Railway Express, among a mix of other railroads. C. T. Wood photograph

2109 4-8-4 on another mail Extra traveling west at California, Missouri, in December 1948. C. T. Wood photograph

2216 is pulling No. 26 "The Texan" into the St. Louis yards in June 1951. The train will soon be backing into Union Station on one of the tracks in the foreground. The 21st Street bridge is in the background. Once a premier train from Texas, "The Texan" is now a coach and café-coach local between Texarkana and St. Louis with no convenient connection off either the Texas & Pacific or the International-Great Northern. The lone passenger car sitting next to No. 26 is where run-through Pullmans to Chicago were normally set out in order to expedite the hand off of these cars between inbound and outbound trains. Joe Collias photograph

A favorite spot in shooting the activities around St. Louis was off the 18th Street bridge, particularly for trains arriving when the morning sun illuminated the train. Joe Collias found 5337 on No. 26 in March 1949.

5340 and 5344 on "The Texan" Grand Avenue Tower southbound to Texarkana in the late 1940s. Texarkana was a massive sorting center for both the Railway Mail Service and Railway Express, and much of the storage mail and express shipments into the southwest were first routed to Texarkana where they were sorted and reloaded to their ultimate destination. This arrangement helps to explain the heavy head-end traffic found on many of the "Iron Mountain" trains. Joe Collias photograph

In 1939–1940, the MoPac shopped 5301, 5302, 5305, and 5307 at Sedalia and gave the four a complete rebuild. This included converting the locomotives from coal to oil burners, larger tenders with 15,000 gallons of water and 5,650 gallons oil capacity, seventy-five-inch drivers, roller bearings, and cast steel pilot beam with a buckeye counterbalance coupler. The rebuilt locomotives were renumbered 5324 through 5327. The boiler pressure was increased to 225 pounds allowing the locomotives to run at 90–100 mph. 5324 was photographed by Joe Collias on No. 19 "The Sunflower" westbound at Kirkwood, Missouri, in August 1949. "The Sunflower" was a St. Louis to Wichita train. The St. Louis to Wichita Pullman (line 3702) was unique in that from August 1948 until August 1955 it was filled by two B&O ten-roomette, five double bedroom cars that were more or less permanently assigned to Missouri Pacific to help equalize car mileage. The two B&O cars, Cascade Drive and Cascade Sound, were painted in Eagle colors.

Pacific-type 6514 is pulling No. 35, the evening commuter local by Lake Junction to Kirkwood in 1940. The locomotive and three cars will tie up at Pacific, Missouri, and return the next morning headed for St. Louis Union Station with suburbanites working downtown. Thirteen new commuter coaches and four baggage coaches were received from American Car & Foundry in 1924 and were used in St. Louis commuter service until the service was discontinued in 1961. Joe Collias photograph

4-6-0 2513 is on the sixty-five-foot Kirkwood turntable after making one of its daily runs on the suburban train down to Union Station in May 1940. 2513 was built for the St. Louis, Iron Mountain & Southern by Rogers and Cooke in 1898.

The Kirkwood commuter trains were discontinued in 1941 with the coach tracks and engine facilities removed in 1945. This area is now the site of a farmer's market. Joe Collias photograph

A work Extra has taken the siding at California, Missouri, while No. 33, a westbound mail and express, passes on the mainline. The horizonal-sheathed wood caboose was unique to the Missouri Pacific being built in 1943. The wood was used to save on steel being used for the war effort. In 1966, the cabooses were shopped at Sedalia with the sides being replaced with steel. Joe Collias photograph

One of the local residents stands in the shade of the depot watching the engineer oiling the running gear on 2-8-0 No. 63 on train No. 457 on August 18, 1948. While this appears to be a mixed train at Delta, Missouri, the timetable shows Delta on this date to have "freight service only." The small consist contains a tank car, coal hopper, and a wood combine being used for LCL and caboose duties. Joe Collias photograph

5535 heads up southbound No. 101, the "Louisiana Sunshine Special," on September 15, 1935, at Monroe, Louisiana. Train 101 carried three twelve-section, one drawing room sleepers from Little Rock to Monroe, St Louis to New Orleans (from No. 1 at Little Rock), and Memphis to Monroe (from No. 201 at Little Rock). A dining parlor car, chair cars, and coaches made up the remainder of the passenger-carrying cars. A. E. Brown photograph

6501 on No. 822 north at Collinston, Louisiana, on September 15, 1951. Train 822 operated from Monroe to Gurdon, connecting with Train 32 at that point. At one time, the RPO line operated as the Gurdon & Natchez, but by this time it had been cut back to operate as the Gurdon & Collinston. The RPO was discontinued in September 1953, and trains 821–822 made their last runs between El Dorado and Collinston/Monroe on April 25, 1955. Collinston, however, retained passenger service on the Little Rock–Alexandria mainline until November 16, 1967. Joe Collias photograph

An I-GN 381 engineer is reading over his train orders before leaving San Antonio on the "Sunshine Special." Before the Missouri Pacific acquired the International-Great Northern in 1926, the 4-6-0s were the largest and most modern locomotives on the Texas railroad's roster. Date or photographer are not available.

An I-GN ten-wheeler is heading up train No. 212 at Corpus Christi, Texas, in August 1950. The mission-style station in the background was built by the San Antonio, Uvalde & Gulf (nicknamed "The Sausage" due to the road's initials).

No. 212 is headed for Odem (on the mainline between Brownsville and Houston) at 11:55 am where it will make a connection with northbound No. 12 "The Pioneer" headed for Houston. Johnnie Gray photographer

Not all Missouri Pacific passengers were provided the luxuries of dining cars, Pullmans, and lounge cars. Here's train No. 832 ready to depart Chester, Illinois, in 1953. Typical of branchline service, the one-car train is a 1200 class 2-8-2 and a single combine 3616. Train 832 ran from Menard to Mount Vernon, Illinois, with stops in between at Chester, Pinckneyville, Tamaroa, and Waltonville. Joe Collias photograph

Missouri Pacific made extensive use of gas-electric motor cars on the lightly patronized passenger routes. As an example, in December 1931 there were nineteen system-wide passenger routes all serviced by various gas-electrics on the MoPac's roster. The 600 was a one-time purchase from General Electric with the car being delivered in 1912; it was unique from the rest of the fleet in having a center entrance and an open-end observation platform on the rear of the carbody. In 1927, 600 was rebuilt by the Sedalia shops with EMC equipment as were the five Sykes gas-electrics the railroad owned. In April 1940, the 600 was found operating two round trips per day on the ten-mile-long branchline between Hoisington and Great Bend, Kansas, but by September that year the branch was timetabled as "freight-only." This view shows 600 outside the St. Louis shops on October 16, 1939, after maintenance. The car was dismantled at Sedalia on September 29, 1948. Joe Collias photograph

Missouri–Illinois 113 has an RPO–baggage car and coach protecting the passenger schedule normally held by M-I's 625 gas electric. The coach has been painted the new Eagle colors. No. 113 was originally Missouri Pacific No. 155, being sold to the M-I in 1952. This photograph was taken in 1952 by Joe Collias at Salem, Illinois.

R. S. Plummer caught T&P Mountain-type 4-8-2 class M-2 907 on film in 1950 pulling train No. 8, which usually was full of head-end cars with a couple of rider cars from Texarkana to Fort Worth. MP train No. 7, which furnished T&P No. 7 its cars, was always a long train. This image shows ten head-end cars with three coaches. At Christmastime, No. 7 became a real problem for Dallas Union Terminal as the train could swell to thirty-five or more cars. That was longer than the tracks in DUT; when this occurred the DUT switcher would have to leave the terminal to meet No. 7 and start working the train from the rear car off the mainline. This photo is south of DUT with the train headed westbound. The water tank at right is the DUT Cadiz Street roundhouse, and the concrete stacks seen above the express boxcars is the Proctor & Gamble soap and detergent plant. The Globe-Union battery plant is on the left.

R. S. Plummer photograph taken at Texarkana in August 1950 off the Broad Street overpass. T&P 4-8-2 907 is leading No. 7 southbound. Missouri Pacific train No. 7, "The Southerner," left St. Louis at 12:45 am with Pullman cars for Hot Springs, Fort Worth, and San Antonio. Extensive switching of head-end cars took place at Little Rock and again at Texarkana. Arriving at Texarkana at 2:30 pm, No. 7 was split into two trains. T&P No. 7, now known as "The Westerner," departed at 3:00 pm for Fort Worth and El Paso. T&P No. 237 followed at 4:15 pm, but at Longview continued over the former I-GN to Palestine, Austin, and San Antonio, with a 7:10 am arrival the following morning.

Texas & Pacific Mountain-type 908, painted in the new Eagle's colors, on train No. 7, "The Westerner," at Texarkana's Union Station. "The Westerner" ran from Fort Worth to St. Louis. T&P engines with pump shields on the pilot deck with the T&P diamonds also had a small brass plates attached to the shields, and the shop forces would stamp the date the valve and piston stems were repacked on the plate. They are visible on the bottom half of the shields. R. S. Plummer photograph

The 702, painted in the streamlined colors of the lightweight Eagle trains (blue and gray with silver border stripes), crosses over the Santa Fe mainline at Tower 55 in Fort Worth with eleven cars of No. 27, the "Louisiana Limited," in 1950. Tower 55 not only controlled the interlocking but also the T&P arrival and departure tracks for the passenger station behind the train. The Pacifics that received Eagle colors were painted at Marshall, Texas, starting in 1947 with the 700, 704, and 706 also receiving large sheet metal along the boiler walk boards and the words "The Eagle." The Pacifics were repainted first and then the Mountain locomotives received the Eagle colors. The Pacifics were done initially because T&P's E-7 diesels were just being delivered in 1947 and 1948 and T&P needed the steam locomotives to supplement the passenger trains until the last of the E-7s were delivered. There still weren't enough E-7s on the property to power the railroad's Eagles, so the Pacifics were assigned to the Louisiana trains south of Shreveport to New Orleans until the T&P's E-8s arrived. R. S. Plummer photograph

Freight Trains in the Steam Locomotive Era

The 445 is captured on film in 1942 on the mainline before going up the branch to switch the Tretolite facility on the Creve Coeur branch. The Creve Petrolite Corporation was headquartered in Webster Groves, Missouri, with their Tretolite Division producing oil field and refinery demulsfiers for resolving water-in-oil emulsions and corrosion inhibitors to protect the piping in oil wells and refineries. The branch broke off the mainline at Lake Junction in Webster Groves and traveled twelve miles to Creve Coeur Lake. Built as the Laclede and Creve Coeur Lake Railroad being finished on July 1, 1881, the branch ran passenger trains to the lake, and early industries included an ice company (located at the lake), a dairy, and a brick manufacturing plant. Wayne Leeman photograph

Engine 134 pulling the second section of train No. 62 north with reefers of Texas grapefruit near C&S Junction, Louisiana, on the Kansas City Southern on March 21, 1937. The head brakeman is standing on the auxiliary tender platform while the train is at speed; it was not uncommon to see trainmen riding on the tops of moving equipment in the steam era. The 134 lasted until the end of steam on the railroad, being scrapped at North Little Rock, Arkansas, on November 30, 1955. A. E. Brown photograph

Train 261 arriving at Branson, Missouri, on September 10, 1947. The lead engine is 1230 followed by a 1522 heavy booster equipped 2-8-2. The Branson depot can be seen in the photograph to the right. This is the foot of the hills south of Reeds Springs, and Lake Taneycomo (White River) is to the left of the photograph. E. G. Baker photographer

MoPac's 1423 is northbound just crossing over the Rock Island mainline at HH Tower in Little Rock, Arkansas. The train's destination is the classification yards in North Little Rock. The low temperature that day in the winter of 1951 was two degrees below zero with four inches of snow on the ground. The photographer captured the scene off the Third Street overpass, undoubtedly a chilly time to wait for a train. Johnnie Gray photograph

The 4000 was the railroad's only mallet locomotive built by Baldwin in 1912 for $36,465; it was purchased for the new Iron Mountain hump yard at Dupo, Illinois. In 1926, the 4000 worked the north hump ten months of the year where only loads were humped, and in December and January it worked the south hump handling mostly empties (the 4000 could shove seventy-five empties on the south hump even at zero temperature). When the USRA 0-8-0s and 1729 were delivered in the late 1920s, the 4000 was temporary retired by the roundhouse in Dupo.

In 1936, it was used on transfer runs between Caronelet Yard and 23rd Street Yard in St. Louis, and in 1937, it was used over the Municipal Bridge between St. Louis and Dupo on transfer runs. It was also used later as a helper engine over Kirkwood Hill from 1940 to 1942 until the 2-10-2s were purchased from the Wabash that same year. L. W. Baldwin, on September 14, 1945, suggested the railroad find a buyer for 4000 due to the condition it was in. The Mechanical Department recommended a sale price of $18,500 and looked at the NdeM as a potential buyer, but there was little interest south of the border. The 4000 is shown here at Dupo on June 22, 1946, shortly before scrapping the following month. Joe Collias photograph

The 416 on No. 457 South at Delta, Missouri, on March 23, 1951. No. 457 originated at Bismarck and ran to Charleston, Missouri. Originally a mixed train, by this date it was timetabled as a "freight only" local. The caboose is from the 1159–1167 series and is uniquely "MoPac," being built at Sedalia in November 1942 as a "coach caboose" with a bay window and LCL area. Joe Collias photograph

The 1711 at Missouri Pacific's Lesperance Street Yard in May 1950 getting ready to depart with a small freight. MacArthur Bridge in background. Alco S-2 9132 is the yard engine. Joe Collias photograph

Johnnie Gray caught this scene on film of Extra 1519 pulling away from the water tank at Sallisaw, Oklahoma, in August 1950.

The Kansas & Arkansas Valley Railway built through Indian Territory starting in 1888 from Van Buren, Arkansas, to Coffeyville, Kansas, and completed the 170-mile railroad in 1890. The property was sold to the Iron Mountain in 1910.

Heavy Mike 1414 pulling a westbound freight over Marshall Road in Webster Groves, Missouri. The photographer, Wayne Leeman, lived a few blocks from this underpass. He could easily hear the Missouri Pacific freights working though the suburbs in time to walk to the tracks and capture the MoPac on film.

The railroad through Webster Groves was finished in May 1853 to a track gauge of 5 feet 6 inches required by the General Laws of the State of Missouri. Once Eads Bridge was built bringing the eastern railroads onto the west side of the Mississippi River into St. Louis and the Union Pacific met the Missouri Pacific at Kansas City, the railroad's board saw the need to change to the standard gauge of 4 feet 8½ inches. This was accomplished in twenty-four hours between St. Louis and Leavenworth, Kansas, on July 18, 1869.

The photograph shows the head-end crew on 1702 at the west end of the helper district in Valley Park, Missouri, waving to the eastbound "Missouri River Eagle." The average train in 1943, using a helper, had between 100 to 120 cars. The small cabin with the stove kept the crew out of the elements while waiting for the next train to assist over the hill.

The these and the next five photographs illustrate the Missouri Pacific's helper service that was used to get the heavy eastbound and westbound St. Louis to Kansas City freight trains over the nearly 1 percent grade over Kirkwood Hill. Helpers going west started at the MoPac's 23rd Street yards as illustrated by the right photograph of 1702 coupled up to 2117 to pick up their westbound freight.

Wayne Leeman photographs

The 1702 class 2-10-2s having just uncoupled at Kirkwood from westbound train No. 75 with 2116 in July 1942.

Engineer John Dauernheim and Fireman Clarence R. Holmes pose with 1702 between helper runs at Valley Park.

Wayne Leeman photographs

Above, the 2-10-2 1711 helper engine cutting off from 4-8-4 2116 on No. 75 for its return trip back to the St. Louis 23rd Street yards in August 1942. To the left is the small freight house and stock chute on a spur just west of the passenger station.

The photograph at the left shows locomotive 1917 coming by the Kirkwood station on the mainline. The track next to the station was for use by the suburban trains. The interlocking tower along with the water tank and turntable for the commuter engine is in the background. Wayne Leeman rode with the head-end crew on the 1917 from Jefferson City to St. Louis in 1940.

Wayne Leeman photographs

The former Wabash 1719 crew is about to couple up to a freight in Kirkwood. The MoPac bought five 2-10-2s from the Wabash in August 1942 that were put to work, as shown here, as helper engines over Kirkwood Hill. Wayne Leeman photograph

Generally, the MoPac used 2-10-2 locomotives as helpers, but the image to the right shows a heavy 2-8-2 pushing a freight near Grants Farm, Missouri, on October 28, 1951. The smoke seen on the extreme left is the head end with Northern-type 2208 heading up the westbound Extra. Joe Collias photograph

Former Wabash 2-10-2 No. 1719 and 4-8-4 2119 traveling through Oakland, Missouri, in the St. Louis suburbs with wartime freight No. 75 pulling 125 cars. After the train gets to the Missouri Pacific yards in downtown St. Louis, the scene will be repeated by the 2-10-2 assisting a westbound freight over the almost 1 percent grade that made up Kirkwood Hill. Photograph by William Barham, July 1943

The wartime volumes increased the traffic on the railroad twice that of the national average between 1941 and 1942, requiring the railroad to need additional locomotives. The Operating Department had hoped to purchase the new FT freight locomotives from EMD, but the War Production Board told the MoPac they had to duplicate a current modern steam locomotive to save on time and materials. Initially, the MP looked at the ATSF 4-8-4s, but they had clearance issues on the MP trackage in Kansas City. The railroad decided to order modified D&RGW M-68 Northerns with a few changes to fit the MP's turntables and clearances. The 2206 was one of fifteen Northern types the Missouri Pacific bought from Baldwin and were delivered in July and August 1943.

The 2206 is leaving St. Louis yards with No. 75 west in 1950 with the 21st Street bridge in the background. MPRR photograph

This photograph shows 2-8-2 1263 is stopped by a signal at Cliff Cave, Missouri, just south of St. Louis at mile pole 12.46 with the Mississippi River being just off to the left of the tree line. The southbound semaphore signal, to the right of the walking brakeman, had two arms—the upper one displayed stop, caution, and clear and controlled movement south through the interlocking limits to the automatic block beyond. The lower arm was fixed and was present to make the signal conform in appearance to a standard two-arm interlocking signal. The brakeman is headed to the line side telephone booth to discuss with the operator at Davis Street Junction Tower when they'll get permission to leave.

Within a few minutes the signal is showing "clear," and the southbound train is departing for Poplar Bluff. CTC was installed in the early 1950s, replacing the remote operations from Davis Street Junction. MPRR photographs

Mountain-type 5336 pulls away from the Gurdon, Arkansas, coaling facilities in 1953 on a southbound Extra. Gurdon was eighty-two miles south of North Little Rock and was a point for taking on coal and water and had a four-stall roundhouse with a seventy-five-foot turntable. Gurdon was the junction of two mainlines: one going southwest to Texarkana and the other southeast through El Dorado into Louisiana.

The 4-8-2 (5335–5344) locomotives were built in two groups, the first being built in 1927 and the second in 1930. When first delivered they were used on the "Sunshine Specials" over the 2 percent grades on the Missouri Division at Gads Hill and Tip Top, Missouri. Once the railroad dieselized their passenger fleet, they were reassigned to the Arkansas and Illinois Divisions hauling freights. Johnnie Gray photograph

Double-headed eastbound freight on the Sedalia Subdivision in the winter of 1944 with two 2100s 4-8-4s. The 2100s were built from the railroad's 1900 2-8-4s purchased in May 1930. The rebuilding of the locomotives took place at the railroad's shops in Sedalia, Missouri. The boilers were lengthened, new cast steel frames were added with seventy-five-inch Box Pok driving wheels, and all the axels rode on Timken roller bearings. The last converted locomotive came out of the shops as "Victory Locomotive No. 2125" on August 20, 1942. MPRR photograph

Local 898's fireman is filling the tender at Scheller, Illinois, on the Mount Vernon Subdivision in 1954. Crew went on duty at Mount Vernon and would wye their engine at Pinckneyville for their return trip back to Mount Vernon. This and the next four pages of images were taken by Wayne Leeman, where he documented the engine facilities, mine switching, and yard at Pinckneyville and rode with the head-end crew on the local in February 1955. The track on hill, behind the water tank, is the MoPac's lead into Orient No. 6 coal mine.

2-8-2 1280 switching at the Pyramid Mine in Pinckneyville, Illinois, in February 1955. A strip-mining operation ceased four years later. The "Route of the Eagles" advertising on the freight cars began in 1948. Wayne Leeman photographs

The 1400 and 1500 2-8-2s were the most numerous class of steam locomotives on the Missouri Pacific roster. The 1444 is shown here at the service track getting sand, water, and coal for the day's work. Blue and gray diesels had started encroaching on the former steam-only territory as indicated by FT 514 on the adjacent track. Photograph taken at Pinckneyville, Illinois, in 1955 by Wayne Leeman.

Water tank, pump house, coaling facilities, and yards at Pinckneyville. Wayne Leeman photograph

Another view of the Pinckneyville yard with hoppers filled with coal the 1280 brought back from Pyramid mine. This area of Missouri Pacific territory was one of the last to dieselize due to the steam locomotives being close to their source of fuel. Wayne Leeman photograph

The local with 1229 spots a Milwaukee boxcar at a feed mill on its run. A MoPac Truck Lines trailer advertising the carrier's "Eagle Service" is at the small freight house for unloading. The image above is the engineer on 1229 watching his fireman pass signals from the brakeman riding on the tender's steps. Wayne Leeman photographs

The 1243 on a transfer run between Dupo and St. Louis on December 9, 1936, in East St. Louis. Note the ice hanging off the Missouri Pacific carload of coal behind the tender. MPRR photograph

In 1949, Wayne Leeman rode a transfer job between St. Louis and Dupo documenting the trainmen and their various duties in the following six photographs. Here the head switchman is seen on a hazy August morning, lining the switch for 1703 to couple into the transfer cut to begin their trip across the Mississippi to Dupo. The 1703 was built by Alco Brooks Locomotive Works in 1916 for the St. Louis, Iron Mountain & Southern as 1503, being renumbered in 1924 to 1703.

The head switchman is passing hand signals to the engineer while riding on the roof of a boxcar. The cut of cars is passing under 12th Street in downtown St. Louis. The transfer job's locomotive is coupling up to the transfer cut destined to Dupo. Wayne Leeman photograph

The 2-10-2 1703 is traveling tender first approaching the Gratiot Street Tower with the transfer back to Dupo. The tracks that run to the left of the tower are called "High Line." On the left-hand side of the tower the train order signal was used by the TRRA mostly to run trains on the TRRA wrong main from Gratiot Street North or from the several different towers located between there and the Merchants Bridge. This track was located between the TRRA mains and the tower. On the right side of the tower are the two main tracks that go across the MacArthur Bridge (owned by the St. Louis Municipal Bridge and Railway).

The additional tracks to the right are auxiliary tracks. The TRRA used these tracks to serve the Ralston Purina Mills until the early 1960s when the mill burned. The TRRA had several regular switch engine jobs that worked the mill. Wayne Leeman photograph

The operator at the Gratiot Street Tower watches Missouri Pacific 1719 pull the transfer out of the St. Louis 12th Street yards to Dupo. His desk shows the track layout that the interlocking plant controls. Although against the rules, during a lull in the railroad activities, he'll check out the Cardinal and Browns baseball scores in the newspaper sitting on top of the control box.

The Gratiot Street Tower was controlled by operators employed by the City of St. Louis and by the St. Louis Municipal Bridge and Railway. They were taken from the TRRA Order of Railway Telegraphers seniority roster and would take a leave of absence from the TRRA to work for the city. Wayne Leeman photograph

On the left, the head switchman is talking to the engineer on the 1703 while the photograph on the right is an engineer, obviously enjoying his work, on another transfer run looking back at the photographer. The train on the right has just crossed the Mississippi Riverbank and is on the Illinois side of the river. Wayne Leeman photographs

This picture was taken going up on the MacArthur Bridge just over 7th Street in St. Louis headed to Dupo. Since there wasn't ample opportunity to turn an engine after it made a trip across the river, it was necessary to run the engine backwards on the return trip to the Dupo yards. Looking through the steelwork for the highway portion of the bridge one can see the bridge work for the south approach that came up out of Lesperance Street that was used by the MoPac, Alton & Southern, and Manufacturers Railway and Terminal Railroad Association of St. Louis. Wayne Leeman photograph

The photographer, while riding 1703's tender, has taken an image of the transfer making its way across MacArthur Bridge, providing a view of the homes and businesses in East St. Louis, Illinois.

The train will travel down the bridge and swing to the right, heading for the Dupo classification yards. In the background, above the railcars, is part of the St. Louis skyline showing though the August haze and humidity. Wayne Leeman photograph

Missouri–Illinois No. 7 (2-8-0) at St. Francois, Missouri, on June 14, 1951. No. 7 is about to pass under Highway 8. Engineer Clayton Huskey is in the window with Leonard (Nook) Stegall standing on the footplate.

The locomotive was built by Alco Richmond in June 1899 for the Mississippi River and Bonne Terre. It was scrapped on March 15, 1952, at Bonne Terre. Joe Collias photograph

June 20, 1951, finds Missouri–Illinois No. 101 on a northbound freight just south of the depot at Elvins, Illinois, and is approaching the town's only street crossing. The train is returning to Rivermines from Derby where it has picked up a block of cars left by a Ste. Genevieve train. The faded signage on the brick building reads, "Opera House." Joe Collias photograph

The M-I No. 104 with a twelve-car freight is northbound near St. Francois, Missouri, on July 7, 1951. The plume of smoke just rounding the curve is No. 11 passing where the depot used to be. The weed-grown track in the lower left is the lead into National Mine. Joe Collias photograph

Missouri–Illinois No. 23 switching the yard at Bonne Terre in June 1949; the NYC gondola is a load of coal headed for the Stewart Coal yard behind the M-I depot. In the background on the left is St. Joseph Lead Company, the chat hill looms in the middle, which were tailings from processing the lead ore, and M-I's concrete coaling facilities are to the right. The switchman riding the car on right is Olin Stewart. Joe Collias photograph

Texas & Pacific 625 on an eastbound train pulling a loaded reefer block east of Longview Junction, Texas, on March 13, 1936. The empty reefers will come back to Longview and travel down the I-GN back to the Rio Grande Valley for loading oranges, lettuce, melons, or grapefruit. The 625 was in the last order of 2-10-4s that incorporated upgrades from the two earlier I-1 orders; a second sand dome would be added over the firebox and one-piece cast cylinders. A. E. Brown photograph

A Missouri Pacific engineer and fireman working on a 4-8-2 Mountain type. Even in the summers, long shirt sleeves were a necessity for the fireman working near the firebox as not to burn their arms from the heat. MPRR photographs

Structures and Classification Yards

View of Missouri Pacific's busy Chouteau Avenue Freight House taken from the MacArthur Bridge in 1942. Note how the boxcars have been precisely lined up to allow the employees to load or unload the lading going from one car to the one adjacent to it. Eads Bridge, spanning the Mississippi River, is in the background. The strings of ART, FGE, and UPFE reefers are placed at what was known as "Produce Row" for unloading. MPRR photograph

Missouri Pacific outbound freight house located at the old Iron Mountain Georgia Street rail yard in Memphis, Tennessee. Missouri Pacific made wide use of billboards to advertise their passenger service. Memphis was the largest cotton market in the world with many loads of cotton originating from this facility. MPRR photograph circa late 1920s

Switching receiving yard 9446 0-6-0 switching the east end of Sargent Yard at Memphis, Tennessee, in 1943. The large building in the upper left was the roundhouse.

For years the 9446 was assigned to the Union Railways of Memphis, which operated all the terminal trackage in Memphis. The locomotive was a product of Baldwin Locomotive Works being built between 1906 and 1910 and was scrapped on February 4, 1948, at North Little Rock. MPRR photograph

These 1570, 1722, 9701 class, and 2101 class locomotives shared the engine facilities at Osawatomie, Kansas, in 1943. In the background is the 1903 thirteen-stall wood-frame roundhouse. Later, eight additional metal stalls were added to enlarge it to twenty-one stalls. Osawatomie engine house and repair shops served three divisions in Kansas: Central Kansas Division, Southern Kansas Division, and Wichita Division. MPRR photograph

This and the next four images were taken at the East Yards in Kansas City, Missouri, in 1942. The largest grain elevator was owned by the Missouri Pacific. The smaller one behind it to the left was the Kansas City Southern's elevator.

To the right of the MP elevator is Elevator "B," then Eagle Elevator, and the last one by the light tower was the Monarch elevator and mill served by the Kansas City Terminal. The base of the light tower is by the hump and in the lower right corner is the turntable pit into the roundhouse. MPRR photograph

Another view of the East Yard looking west with the large Continental Elevator, built by the Missouri Pacific, that could store seven million bushels of grain. In 1941, the elevator's four storage tracks held a total of 344 cars and the four receiving tracks held 148 cars. Note the large Missouri Pacific buzz saw painted on the side. MPRR photograph

An 0-8-0 pulls a cut of cars out of one of the bowl tracks in East Yard. Elevator "B" sits in the background to the right and the Missouri Pacific elevator is on the extreme left. MPRR photograph

MP USRA 0-8-0 classifying a cut of cars. There was no radio communication in this era with all the switching communicated by hand signals or using lanterns at night.

The Northern Pacific forty-foot steel underframe reefer that 9778 is coupled on to was part of a group of cars built by the NP railroad's South Tacoma shops in 1931–1932: 9000–90799 and 90800–90999. MPRR photograph

MP switcher 9778 adds a Magor caboose to the end of a freight that's ready to depart the East Yard at Kansas City. Note the large MoPac emblem displayed on the railroad's elevator. The skyline is populated with pigeons feeding on the spent grain leaking from the freight cars waiting to be unloaded. MPRR photograph

Due to the growth of traffic moving on the Missouri Pacific during the war, the railroad found themselves short of locomotives and cabooses. To partly help alleviate the caboose shortage, ninety NOT&M forty-foot single-sheathed boxcars were shopped in 1942 with sixty at North Little Rock and thirty converted at Sedalia. One example, NOT&M 2895, right, sits on the third track from the right. MPRR photographs

The lines of tank cars in "F" Yard at Dupo are a grim reminder that German U-boats were taking a toll on the oil tankers traveling up and down the East Coast, sinking 22 percent of the fuel tankers in less than seven months in 1942. In addition to the loss due to the submarines, most available ships were diverted to carry fuel to Great Britain. By 1943, the railroads were hauling a million barrels of petroleum a day.

Scene looking into the engine facilities at Dupo, Illinois, in the 1930s. In the upper right is the concrete roundhouse with the turntable to the right. The foreground had loads of company coal destined for the coaling station. On the lower right is the ash pits and hoist frames for dumping ash from the fireboxes into the gondolas. MPRR photograph

9771 pulls a track towards one of the leads to start classifying the cars into outbound blocks. Image taken at North Little Rock, Arkansas, in 1942.

9771 was a USRA 0-8-0 copy built by Lima in 1929. It was a slightly different design than the original USRA locomotives as it had a cast frame which decreased maintenance versus having a frame that was bolted together on the original locomotives. MPRR photograph

Two auxiliary tenders are in the foreground at the large shop complex at North Little Rock. Locomotives include 2-8-0s 14 and 135 with a 6400 class Pacific on the next lead over. The auxiliary tenders were modified from MoPac-owned tank cars, leased cars from General American, and private tank cars that had been purchased from their owners after the cars had been involved in accidents on the railroad. The MoPac first used one of these tenders in the summer of 1926 testing on freights between St. Louis and Kansas City. The railroad saw benefits in both fuel savings and the elimination of some water stops. By May 1940, the railroad had eighty-seven auxiliary tenders in service averaging more than 6,000 miles a month. Savings in 1940 by using these cars was estimated at $8,000. As the railroad dieselized, the leased GATX cars were shopped at North Little Rock with the domes being refitted and returned. The remaining cars were placed in Maintenance of Way service. MPRR photograph circa 1941

View looking down the service tracks at the joint MoPac and Texas & Pacific facilities in Texarkana in 1948. The structure on the left is the cinder conveyer used to move the dumped cinders from the locomotive firebox to a gondola. Behind 1417's tender is the Roberts and Schaefer coaling device, and the sand tower and water column are to the left of 902. R. S. Plummer photograph

Passenger Trains in the Diesel Era

Besides puling the "Colorado Eagle," the EMD E-6s were used on some of the heavyweight trains. No. 15 westbound "Royal Gorge" is shown with one of the diesel sets. The postwar version of the "Royal Gorge" was inaugurated on June 2, 1946, as secondary trains 15 and 16 on the St. Louis–Kansas City–Pueblo line. The train initially carried a through Pullman to San Francisco that was handed off to the Rio Grande at Pueblo and to the Western Pacific at Salt Lake City. Starting in 1948, for a brief period, the entire train ran through to San Francisco, but ridership never matched expectations and through service disappeared in 1950. This scenic photograph was taken on May 28, 1947, near Gasconade, Missouri. MPRR photograph

The "Colorado Eagle's" first run was on June 21, 1942. Wayne Leeman was out the following month in west St. Louis County to capture the railroad's newest train on film. The train ran as train numbers 11 and 12, replacing the "Scenic Limited" that was put in service in 1915.

No. 11, the “Colorado Eagle,” departs St. Louis Union Station on August 29, 1946, at 4:00 pm with a mix of cars. The consist shows the original Budd equipment ordered for the train, a coach from the original 1940 Eagle, a pair of 6-6-4 sleepers, and a modernized Tower series heavyweight Pullman being used as an overflow sleeper. On either side of the stainless-steel Eagle, the nose is lettered for the two railroads handling the train—“Missouri Pacific” on the left and “Rio Grande” on the right. The two early Alton EA units were backing into the train shed to couple up to the “Ann Rutledge” for a 4:30 pm departure to Chicago. MPRR photograph

With the new streamliner "The Eagle" operating between St. Louis and Omaha, the railroad purchased Motorailer 670 from American Car & Foundry in January 1941 to connect with the new train at Union, Nebraska, and run to the state capital at Lincoln. Due to the demands of World War II, it was not placed in service until September 5, 1942. Unfortunately, 670 was plagued with mechanical problems with both the Waukesha engines and transmissions. After meetings with AC&F and several studies, it was decided to repower 670 with Cummins diesels and Allison transmissions and the final run between Lincoln and Union was held on July 10, 1954. The car was sent to Sedalia for the new upgrades. After being released by the Sedalia shops in early 1955, it was assigned to the Delta Eagle running between Helena and McGehee, Arkansas. The mechanical issues continued to plague 670 with clutch issues and wear to the diesel's pistons due to roadbed dirt with the railroad sending it to the North Little Rock shops for maintenance.
By 1960, the railroad made the decision to retire 670 and try and sell it. Not surprising, given the poor track record of mechanical issues, there were no buyers and 670 was scrapped at North Little Rock. MPRR photograph taken at St. Louis in 1942.

When “The Eagle” was put in service on March 10, 1940, the maximum speed on the Eastern Division was ninety mph on tangent track and eighty on non-restricted curves. Here’s the eastbound train No. 6 at Jefferson City, Missouri, on May 28, 1943, now renamed “Missouri River Eagle” (after the river shown in the background) loading wartime passengers. Departure will be 3:30 pm.

Note the “Eagle-ized” heavyweight coach next to the RPO–baggage car. Due to the increased popularity and ridership of the new streamliner, in September 1940 four 1915-era heavyweight coaches were shopped at Sedalia and rebuilt into Deluxe Coaches (6326, 6327, 6328 and 6329) for the new train. To blend in with the lightweight cars besides the Eagle colors, updates also included thermopane windows, updated interior to match the lightweight cars, and one vestibule removed at the end of the car to enlarge the men’s and women’s restroom areas with lounge sections. MPRR photograph

Eastbound train No. 6, the "Missouri River Eagle," loading passengers and mail at Jefferson City in April 1958. By this time, MoPac's first streamliner was carrying a substantial volume of head-end traffic, as evidenced by the Santa Fe baggage car ahead of the RPO. The RPO itself is unusual; the normally assigned car was a combination baggage–mail car with a thirty-foot RPO compartment, but today's train has one of the Sedalia-built sixty-foot postal cars and a separate heavyweight baggage car. Note the RPO clerk "robbing the box," retrieving mail from the trackside mailbox. In September 1958, the entire RPO operation on train 6 was rescheduled to operate on train No. 14.

The passengers sitting on the left side of the train will have a great view of the Missouri River for the next few miles before the mainline cuts across the Missouri landscape to St. Louis. Wayne Leeman photograph

Missouri Pacific's first E-7A and B locomotive 7004 was delivered in September 1945 with the A unit being the only MoPac E-7 locomotive having the small number boards. The 7004 is shown here, December 5, 1945, heading up the eastbound "Colorado Eagle" out of the downtown St. Louis yards. The consist is somewhat unusual with two RPO–baggage cars (one being quite grimy) and an original "Eagle" car (five cars back from the "B" unit) added to the train as noted by the higher gray band than the rest of the cars. The 21st Street bridge, a favorite spot for photographing trains, is in the background. MPRR photograph

Two new Alco PA-2s heading up the “Colorado Eagle” No. 12 through California, Missouri, in almost a Christmas card scene, winter 1948. C. T. Wood photograph

A lone EMD E-7A 7014 is on train No. 116 North at 6:60 am in Monroe, Louisiana, in June 1955. At this time, No. 116 was a New Orleans to Little Rock train, overnight between New Orleans and Monroe, leaving New Orleans at 7:50 pm and arriving in Little Rock at 12:15 pm. The train carried a heavyweight Pullman (ten section—one drawing room, two compartment configuration) and coaches between New Orleans and Little Rock. A grill coach was carried between Monroe and Little Rock. MPRR photograph

Train 238 with 8008 in the lead negotiates the complicated trackage (crossing both the Cotton Belt and the Kansas City Southern) just prior to arrival at Texarkana Union Station. In addition to head-end traffic, the train carried a heavyweight sleeper from San Antonio to Little Rock, a dining–lounge car between Longview and St. Louis, and, of course, coaches between San Antonio and St. Louis. At Texarkana, T&P train 238 from San Antonio arrived at 11:30 am and T&P train 8 from Fort Worth arrived at 11:45 am. Both were consolidated into MP train No. 8 in time for a 12:05 pm departure for Little Rock and St. Louis. Although a "name train," "The Southerner," this train always carried a heavy volume of mail and express traffic. Photograph taken September 16, 1951, by R. S. Plummer.

The T&P delivered two trains to Texarkana which were combined into MP train No. 8 for movement north to St. Louis. This December 1951 view is not identified but would be either T&P train No. 8 from Fort Worth or T&P train No. 238 from San Antonio (delivered by the I-GN to T&P at Longview).

The T&P Swamp Holly orange and black, station switcher, is handling a GM&O baggage car. The switch crew on the west end of the station was supplied by the T&P, while the switch crew at the east end of the station was a MoPac crew. R. S. Plummer photograph

A single PA-1 8003 pulls a northbound “Texas Eagle” No. 22 at Austin, Texas, in the summer of 1951. Note the absence of any head-end traffic; at this time Nos. 21 and 22 did not carry any checked baggage or remains, and No. 22 did not pick up a working RPO until it reached Palestine. The dormitory–coach behind the locomotive was probably used to help muffle the engine sound for the revenue passengers. R. S. Plummer photograph

The photographer is standing behind the engineer on the head end of southbound No. 3 through Piedmont, Missouri, in April 1953 in a MoPac "PA" unit. The native stone depot would be to the right of the northbound No. 4 with a pair of PAs. The photographer's train has just passed the depot on his left where Tuffy Graham was the MoPac agent. No. 4, who just covered eighty-five miles through the Ozark mountains, was due to Piedmont at 1:48 pm and No. 3 at 1:55 pm.

No. 4 carried a Little Rock to St. Louis café–parlor–coach as well as Fort Worth to St. Louis coaches. No. 3 had an identical consist with both trains carrying St. Louis and Little Rock RPO cars. Moss Tie yard and Worley Tie and Timber yard are on the right adjacent to the steam engine. Several Maintenance of Way outfit cars are seen sitting on the right side of the photograph. Wayne Leeman photograph

Santa Fe F-7 326 on the Texas Chief and PA-1 MP 8012 on the Texas Eagle make a great comparison between the Alco and EMD locomotive designs. The Houston passenger station and offices are just to the left in the image. The area where the locomotives are is the current site of the Houston Astros baseball stadium that opened in 2000. R. S. Plummer photograph

The 8029 PA-2 departing Houston on July 12, 1953, with No. 121, the "Valley Eagle," at 10:45 am. Brownsville, Texas, is 372 miles south. The "Valley Eagle" went in service on October 31, 1948. The Railway Express building and various REA delivery trucks are directly behind 8029. R. S. Plummer photograph

A single FA-2 with a steam generator pulls an Extra mail and express train on the Kansas City Terminal tracks towards Union Station a few weeks before Christmas in 1952. These trains were usually run as second sections of trains 7, 8, 9, or 10 consisting of an overflow of second- and third-class mail and express due to the holidays. The eleven-car consist is a mix of ownership including a REA reefer, two Pennsylvania X29 express boxcars, two MoPac fifty-foot boxcars painted in Eagle colors (one as the last car on the train), and a heavyweight rider coach for the crew. The tracks above the cut to the upper left belong to the KCT.

Missouri Pacific owned a large Alco FA fleet with numbers 361–373, 387–392 being equipped with steam generators for dual freight and passenger service. Joe Collias photograph

Train No. 231 "Southern Scenic" makes a stop at Bergman, Arkansas, in October 1958 to unload express packages. The GP7 4253 was originally purchased for the St. Louis, Brownsville & Mexico in 1953 equipped with a steam generator for passenger service. The small consist consisted of two "Eagle-ized" heavyweight cars, an RPO–baggage, and a coach. 1950s television talk host Arthur Godfrey is featured on the side of the REA truck promoting Chase and Sanborn coffee. Wayne Leeman photograph

Having just changed crews, No. 37 is on the north end of the yard heading south through Poplar Bluff in April 1962 with an E-7 in the point. No. 37 was a first-class mail train at one time known as Advance No. 7 ("The Southerner"). Coaches on northbound No. 32, the "Sunshine Special," came back deadheaded on this train when the only "Sunshine" remaining was the northbound train from Texarkana to St. Louis. The yard office is the white building behind the row of automobiles; the small, elevated building to the right is the gate watchman's shanty to guard the 4th Street crossing in the yard. The shanty was normally manned by disabled railroaders with some degree of restrictions. The large building to the right with the Purina Mills checkerboard is Forbridge Feed & Produce. The 1910 passenger station would be behind the photographer. Wayne Leeman photograph

After the Pullman divestiture of sleeping cars to the railroads, two eight-section, one drawing room-two compartment sleepers were acquired by Missouri Pacific subsidiaries. "Paris Gibson," pictured above, was acquired by StLB&M and "Marcus Daly" was acquired by I-GN. "Paris Gibson" was painted in Eagle colors in April 1948 and "Marcus Daly" two months later. Both cars were then assigned to Pullman line 3500, operating between St. Louis and El Dorado, Arkansas. Pullman line 3500 enjoyed extensive utilization by employees and officers of Lion Oil Company, which was headquartered in El Dorado. These cars normally moved in trains 31 and 32 between St. Louis, Little Rock, and Gurdon, where they were transferred to and from the Gurdon–El Dorado trains. When train 31 was consolidated into trains 1 and 21 on August 15, 1955, it was necessary to convert this Pullman line to lightweight cars because of MoPac's reluctance to handle heavyweight cars in the "Texas Eagle" consists. The two cars were reassigned to Pullman line 3641, Houston–Brownsville, beginning in that service on August 25, 1955. This Rio Grande valley assignment lasted until November 22, 1957, when the two cars became unassigned, available for use anywhere in the Pullman Company pool. "Paris Gibson" was removed from Pullman lease in November 1959, stored at Sedalia, and sold for scrap in November 1962. MPRR photograph

Unfortunately, here's what the 1948 fourteen-car streamlined "Texas Eagle" trains morphed into by the end of MoPac's passenger service. This train only operated as a turn from St. Louis to Texarkana with the Texas portion of the train route (running from Texarkana to Laredo) being eliminated earlier on September 21, 1970.

Missouri Pacific's E-8 and a grill coach is shown here arriving in Poplar Bluff, Missouri (station) on March 17, 1971. No. 38 was originally delivered in 1950 as 7018 and featured a nine-inch stainless steel Eagle on its nose. Amtrak is just forty-four days away. Wayne Leeman photograph

Missouri–Illinois gas-electric 625 and its RPO–baggage trailer are at Branch Junction, Illinois, having just gotten off the Illinois Central's mainline. The M-I used 2.4 miles of the IC's mainline through Central City and Centralia getting on the IC at IC Junction and off at Branch Junction. This is train No. 2 running northbound from Kellogg to Salem, Illinois, daily except Sundays. The 625 was originally numbered 606 and delivered as a Sykes gas-electric in 1925. Due to the mechanical issues with the original power plant car, it and four other Sykes cars were rebuilt in 1927 and 1928 with EMC equipment at Sedalia and renumbered. Joe Collias captured No. 2 on film on July 21, 1951.

T&P F-7A 1581 and E-8A 2013 leads No. 15 "The Texan" over the Belt Junction diamond near Dallas, Texas, on June 22, 1952. The 1581 was one of four F-7A units regeared for passenger service. It was later repainted in the passenger scheme by the railroad. The F-7As weren't equipped with steam generators but did have steam and signal plumbing to match with the E units, in the event they were in the trailing position. If trains using F units were long, the railroad would add a steam generator car.

The interlocking controlling Belt Junction was Tower 118, placed in service on September 24, 1926. It was built and manned by the SP–T&NO. The tower had a forty-lever electric machine controlling Metzger, Belt Junction, the original H&TC mainline with the Belt Line, and both ends of Fox Siding. R. S. Plummer photograph

Freight Trains in the Diesel Era

The 9102 crew stops for a photograph for the company photographer in 1942 just north of Rutger Street in the Lesperance Street Yard. The Alco HH1000, built in September 1939, was one of the first diesels switchers purchased by the MoPac. It was assigned to the St. Louis Terminal for a couple reasons: first to help reduce smoke in the downtown St. Louis yards due to a smoke ordnance passed by the city, and secondly to keep the new motive power in the shadows of the MoPac's general office where the Mechanical Department could assess its qualities against the Baldwin and EMC diesel switchers also on the property.

This was the only turbocharged locomotive the railroad owned until the second generation EMD locomotives were delivered in the 1960s. Alco only built thirty-four of the HH1000s. The tall structure behind the locomotive is the south approach on MacArthur Bridge. MPRR photograph

A view of the third-shift switcher spotting a UPFE reefer at Produce Row in St. Louis for unloading. This image reinforces railroading is a continuous "twenty-four hours a day, seven days a week" industry with most holidays treated as normal workdays. At the time of the photograph, there were no radio communications between the switchmen and engineer—all signals being conveyed by lanterns after sunset. The black switcher 9129 was an Alco S-2 built in June 1949. Wayne Leeman photograph

These two photographs, taken in December 1947, show Alco S-2 9113 working on the Creve Coeur Branch at Lake Junction, Missouri. The covered and open-top hoppers are on the Breckenridge Concrete Company track. These will be pushed over an unloading trestle, one at a time, by an old Mack truck weighted with concrete slabs. After unloading, the cars rolled back through a rubber switch to an “empty track.” Hoppers would be loaded with sand and gravel from the Meramec River at Glencoe on the MoPac’s old main line. The five tank cars are for loading at Tretolite with specialty chemicals destined for the used in the oilfields. A St. Louis Public Service PCC car can be seen crossing over the MoPac’s mainline. The small Lake Junction depot was used as a commuter stop. Wayne Leeman photographs

9113
MISSOURI PACIFIC
LINES

The Texas Lines purchased six of the Alco S-2 models with five going to the International–Great Northern (9156–9158 and 9168–9169) and one to the St. Louis, Brownsville & Mexico (9159). While the MoPac upper line switchers were painted in the MoPac's traditional steam locomotive black, their Alco, Baldwin, and EMD counterparts in Texas received additional orange striping and pilots for increased visibility.

The 9158 with two T&P stockcar was photographed by R. S. Plummer at Palestine, Texas, 1949.

San Antonia, Uvalde & Gulf Baldwin VO660 switcher at the Brownsville passenger station on March 19, 1947. The locomotive was built in June 1941. Ralph Carlson photograph

FT 511 and its "B" unit passes under a signal bridge in Newport, Arkansas, on a northbound freight in 1947. MPRR photograph

During the floods of 1927 that April, rains soaked the Arkansas River Valley and caused the Arkansas River to flood. As the floodwaters reached Little Rock, the Missouri Pacific placed coal cars on the old bridge for ballast, but the bridge washed away on April 21. The railroad took over the ownership of the bridge and pledged to rebuild it.

The new bridge opened on February 2, 1929. Here, a northbound freight crosses the Baring Cross Bridge into the North Little Rock classification yards in 1956. MPRR photograph

Service track at North Little Rock, Arkansas, in 1955. The Missouri Pacific company photographer won first place in a contest for this image, among other Class 1 railroad photographers. MPRR photograph

The FA-2 No. 334 westbound at Webster Groves, Missouri, in 1953 with an all-Alco ABA lash-up. As the MoPac dieselized, being able to add multiple diesel units on a freight negated the need for helpers over Kirkwood Hill. Wayne Leeman photograph

Topping a grade on the International–Great Northern just north of Valley Junction, FA-1 314 leads an all-Alco ABBA lash-up through Hearne, Texas. Joe Collias photograph

The Alco FA-2 379 and three "B" units lead a northbound freight through Cricket, Arkansas, in October 1958. The depot can be seen behind the boxcar in the consist and the barn structure behind the depot was used to store cans for a local canning factory.

The Missouri Pacific's White River Division, built in 1904, had five tunnels and fifty-two bridges to illustrate challenges building a railroad through this part of the country. Wayne Leeman photograph

Missouri Pacific ordered eight (numbered 4104–4111) of the BL-2 diesels (BL denoted Branch Line service) from EMC at a total cost of $1,054,168 in 1948. The 1,500 horsepower locomotives lasted until 1962 when they were traded in on the GP-18s. Only nine railroads purchased the type with only fifty-nine total being built.

The 4106 and 4104 at Hutchinson, Kansas, on June 10, 1948. The imposing station in the background is Santa Fe's station built in 1897. To modernize the railroad, it was replaced in 1954 by a one-story contemporary brick depot. MPRR photographs

MISSOURI PACIFIC LINES

The F-7 number 571 with eighty-nine cars leaves with a northbound train at Malvern, Arkansas, in May 1951. The tank cars behind the locomotives originated from the oil fields around Smackover and El Dorado, Arkansas, which were some of the largest oil deposits discovered in 1922. MPRR photograph

On an overcast day in 1952, Joe Collias caught a pair of EMD F-3s on northbound freight at Ware, Illinois. The Mississippi River is just behind the tree line on the left.

The 521 and 556 at Palestine, Texas, in 1949. Palestine was the home of the International & Great Northern. The railroad's 1892 passenger station sits behind the three F-units, and the Railway Express Agency building is the large building to the right. R. S. Plummer photograph

Waiting a few minutes, R. S. Plummer shot 521 again departing the Palestine yards in 1949.

Missouri Pacific ordered eight “A” and four “B” units (201–208) of the DR-4-4-1500 locomotives each having 1,500 horsepower at a cost of $1,758,247, being delivered at St. Louis, Missouri. Lightweight of the three-unit set was 703,000 pounds. The 208 with a B and A unit at Jefferson City arrived off River Subdivision on January 30, 1953. Baldwin’s pneumatic MU system was non-standard that didn’t allow them to be run with EMD or ALCO diesels, ensuring they ran as all Baldwin A-B-A sets. Other assignments were out of Texarkana. The fleet was retired in September 1957 and scrapped in September 1959 with their trucks and traction motors being used on the Alco RS-11s. These two photographs show one of the trios at Jefferson City, Missouri, idling for a crew change. Joe Collias photographs

SOUTHERN
LINES
PACIFIC

The EMD GP-7s 4179 and 4180 are newly delivered from EMD seen here at Webb City, Missouri, on May 27, 1951. The coal hopper and portable conveyer would soon be retired as they were no longer required to service steam locomotives with the new motive power assignments. Joe Collias photograph

The GP7 4189 on train No. 458 North on the Belmont Subdivision at Delta, Missouri, waiting to cross the Cotton Belt tracks on its way to Bismarck, Missouri. The freight with the FAs is southbound to Poplar Bluff. When this image was taken, the MoPac had five trains to Little Rock and two to Paragould. The MoPac shared Cotton Belt's tracks through a trackage rights agreement dated September 4, 1904. The St. Louis, Iron Mountain & Southern ran over the SSW from Illmo Junction, Missouri, to Paragould, Arkansas. Both roads at the time were owned by Jay Gould. The 1904 costs in the agreement to the Iron Mountain was $6,442 monthly.

As the two railroads were under separate ownership (MoPac and Southern Pacific), the relationship between the two Operating Departments on "whose trains got priority over the other" was generally in debate until the UP purchased the SP. Wayne Leeman photograph, May 1956

The GP-7 4189 on train No. 458 north on Belmont Subdivision. The train is coming out of the Glen Allen tunnel near Lutesville, Missouri, in May 1956.

The 120-mile line was completed from Bismarck to Belmont, Missouri, on August 14, 1869, and crossed the Mississippi River via a railroad car ferry to Columbus, Kentucky, where it connected with the Mobile & Ohio Railroad. The Iron Mountain initially built the branch to five-foot gauge, allowing interchange of cars with the M&O and other southern railroads that were using that track standard. The branch's gauge was changed to 4 feet 8½ inches on June 28, 1879, which was the railroad's new standard gauge for the Missouri Pacific and Iron Mountain. The car ferry was used until late 1911. Wayne Leeman photograph

A single GP-7 4234 on a local freight west of Mount Vernon, Illinois, in 1953. Joe Collias photograph

Looking west at the GP-7 4282 switching the Scott County Milling Company at Sikeston, Missouri, on a cold March 3 in 1960 indicated by the patches of snow. The 4282 was on local 700/701, which originated at Poplar Bluff as a "turn" to Charleston six days a week. The mill was located three blocks west of MoPac's Sikeston depot, which is now a museum.

The Gristo Feed signs located at the top were lit up at night and could be seen for miles. The Bowman family ran the elevator along with other family members. The Katy boxcar shown on the left is sitting on the mainline.

Wayne Leeman took this and the next four images while documenting the various locals switching the mill.

Train No. 451 with GP9 4333 shown with wood caboose 873 switching the four tracks into Scott County Milling in May 1956. In 1960, business from the elevator was booming with corn and wheat shipments, with rail being the largest mode of transportation. Wayne Leeman photographs

March 16, 1960, finds GP9 4335 on Local 701 westbound headed towards Poplar Bluff. The group of forty-foot boxcars on the head end could be loads of grain that were picked up at the elevators at Charleston, Bertrand, and Sikeston or empty boxcars were pulled from Malone and Hyde Grocery in Sikeston. The train is coming across the Frisco main line with the milling company being in the northeast quad of the two railroads. The large smokestack with SCMCO was used for burning the corn cobs that came from shelling corn. To the right is an Illinois Central boxcar sitting on McKnight-Keaton Wholesale Grocery Warehouse tracks. Wayne Leeman photograph

Missouri Pacific agency in Sikeston, Missouri, in March 1960. The individual sitting at the cashier's desk is Clarence W. Elliott, who was working one of the telegrapher jobs. J. E. Robinson was the Star Agent, J. C. Whittaker was the daylight telegrapher, Mose Waller was the cashier, Doc Roberts was the OS&D clerk, and Ken Kiersey was the general clerk. The desk against the wall by the windows is where he'd normally be working, but the photographer had him move so more of the depot's interior office could be shown. The two upright levers mounted on the desk connected to the train order signals outside the bay window; the two wooden forks next to the 1960 calendar were used to hand up train orders to the passing train crews. Also shown on the operator's desk is a telegraph sounder, key, and dispatcher's telephone—all part of the period communication tools to keeping the railroad moving.

Mr. Elliott started with the Missouri Pacific on February 7, 1920, and retired in May 1960. He and many other "old head" employees retired in 1960 as if they'd continued working, they would lose their government pension from serving in the First World War. Wayne Leeman photograph

The Alco RS-3 4501 and 4518 at Mount Vernon, Illinois, heading southbound in August 1956. The loads of round wood are destined to the West Virginia Pulp & Paper Company's paper mill located in Wickliffe, Kentucky. Wayne Leeman photograph

R. S. Plummer caught EMD F-7 551 on southbound freight at Texarkana, Texas, on October 7, 1956. The joint MP-TP engine facilities sit to the right.

All the diesels in this photograph are of Texas & Pacific ownership—two sets of F units, an E unit, and NW-2 switchers working the yard and passenger station tracks. The 1518 is from the first group of F-7s delivered to T&P. Texarkana, Marshall, and Longview were all "hot spots" for freight and passenger trains and received more than anywhere else on the railroad. The 1518's southbound freight is headed for Longview or Fort Worth. Photograph taken in Texarkana 1956.

The T&P's last steam run was No. 396 a 4-6-0 on November 27, 1951, between Shreveport and Alexandria, Louisiana, beating the MoPac by almost four years in retiring their steam locomotives and fully dieselizing. R. S. Plummer photograph

The small 44-ton diesels were successful as internal correspondence from May 20, 1941, noted that in July 1941 additional models of the little diesels were due to arrive. Three more Whitcombs were destined for assignments at Lake Charles and Monroe, Louisiana, and the run between Jackson and Delta, Missouri. Davenport Besler was to deliver three of their 44 tonners in August 1941—two for Joplin, Missouri, and one for Arkansas City, Kansas. General Electric was to receive their 44 tonner in August 1941 to work between Charleston and Sikeston, Missouri.

The 804 is at Lincoln, Nebraska, on September 12, 1954. Lou Schmitt photograph

The Whitcomb 804 was built in November 1940 having 386 horsepower and costing the railroad exactly $29,592. The little 44 tonners were popular due to their weight on drivers, meaning no fireman was required which saved the railroad the costs of an extra crewman. The 804 is shown here switching the MoPac yards in Lincoln, Nebraska, on September 12, 1954. Lou Schmitt photograph

The GP-18 535 leads two companion GP-18s on train No. 75 through Webster Groves, Missouri, in October 1963. The second unit 522, still in its original Eagle Blue and Gray colors, was delivered as 4818 with a high nose that was later shopped by the railroad to match the 535 low nose built in May 1962. The trailing unit was 550, another low nose GP-18 that was built in January 1963. Missouri Pacific was the largest buyer of the GP-18 with 146 on the roster; 36 percent of the total built for all railroads. This large purchase allowed for the retirement of the aging Alco FA fleet and converting the railroad to an all EMD-powered railroad—simplifying parts distribution and maintenance for the Mechanical Department. Wayne Leeman photograph

Miscellaneous MoPac

"Dinner to go." In 1949, an around-the-clock railroad beanery opened that served just one group of customers—train crews passing through Illmo, Missouri, on their way to and from Dupo, Poplar Bluff, or Paragould. The trainman's union agreement dictated that the railroad pick up the meal costs. The trainmen would write down their orders at the origin station and give it to the telegrapher, who then relayed the orders to the café staff. A specially built stand was constructed where the train would slow down to 15–20 mph, pick up the meals off the hook, and resume to their normal speed. Here, the photographer captures the head brakeman grabbing the meals for the head-end crew while the conductor and brakeman in the caboose will get theirs in the same manner. The only known mishap was when a hobo riding the rails fifteen cars ahead of the caboose got the food before the conductor did. Wayne Leeman photograph

While not the highly visible occupation like an engineer or conductor, without the agents, operators, and clerks on the railroad no trains would move, no railcars would get to customers, and no revenue would be coming into the coffers. The clerk is shown placing a placard "Do Not Hump" on MP 94325. The railroad issued a 218-page book on April 1, 1927, titled *Rules and Instructions to Station Agents* with 1,591 rules to be governed in running an agency. In addition to knowing the station rules, the employees had to be knowledgeable of the operating rules. MPRR photograph

Crew Caller Donald I. Farrar Sr. overlooking the various yard, transfer jobs, and train assignments. This job was complex in they had to understand the labor agreements in filling positions and call the men based on their seniority date and qualifications, if they'd had their required rest period, and what division of the railroad they could work. Some assigned jobs would be "show up" jobs while others might require a call. Even though the job might be assigned, the start time could vary daily, and the crew would have to be called. For vacancies due to sickness or vacation, an employee would have to be called to fill the vacancy. The caller would go down the extra board in order and call each person in turn to fill the next vacancy. If the extra board was exhausted, the vacancy would have to be filled by calling persons in seniority order. Depending on the job to be filled and local agreements, it could be necessary to take an individual off their assigned job and move them to the vacancy. This exercise could be quite involved. Figuring in rest days on current assignments, length of an employee's current rest period, and if they could be "forced" to protect the vacancy in question. Also, the question of being qualified was a concern—was a hostler or fireman qualified to be an engineer. Calling the wrong trainman would result in a time claim against the railroad.

The office was located at 3001 Chouteau Avenue in St. Louis. Wayne Leeman photograph

Nelson Collins is shown here copying train orders from the dispatcher. He was one of the last operators manning the Kirkwood Tower. He had originally started on the Missouri Division, eventually transferring to the Eastern Division. At one time he worked a position called the "long swing job," which covered the agent's job at Kirkwood on Sundays, the operator job at Warrensburg on Mondays working from 12:01 pm to 9:00 pm, the operator's job at Jefferson City on Tuesdays from 11:00 pm to 7:00 am, the operator's job at St. Louis 23rd Street on Wednesdays from 11:00 pm to 7:00 am, and finally the operator's position at St. Louis Grand Avenue Tower on Thursdays from 11:00 pm to 7:00 am. His favorite line with regards to any question about working at one of these jobs was, "I only work here one day a week." It was probably a valid answer until the trainmaster, Joe Mellon in St. Louis, told him he would fire him if he heard it again. After Nelson retired in the early 1970s, he played horn in a band on excursion ships out of Miami. This and following page photographs by Wayne Leeman in April 1964.

The image to the left shows Collins putting out the train order forks and below the head brakeman is standing in the doorway of No. 15, the E-7, ready to pick up the head end's set off the top fork.

The Kirkwood interlocking tower was opened on April 12, 1910, having thirty-two levers that controlled a three-track mainline, storage, and yards tracks in Kirkwood and the Carondelet branch. It was closed in 1964 and demolished on April 9, 1964. Wayne Leeman photographs

The Grand Avenue Tower was opened in 1886 and closed 1987. A ninety-five-lever machine controlled the four hundred movements per day. On the west side of the tower was the Frisco, Missouri Pacific and Wabash and on the east the TRRA, Frisco, Missouri Pacific and Wabash. Until 1955 when a GRS NX system was installed, two employees worked each shift—one a leverman to make the lineups of each movement and a clerk to answer the phone and keep a written record of each movement. Wayne Leeman photograph

Wayne Leeman documented a switch being replaced at the Grand Avenue Tower in St. Louis by two 110-ton cranes. The large "I" beam kept the rails and ties from shifting while it was being put in place.

Having pulled away from the Ivory Street incline, the *Willard V. King* traveled upriver a short distance to the incline at East Carondelet, Illinois. Photograph taken in July 1938.
MPRR photograph

The image on the left shows the ferry pulling away from the incline on the Missouri side of the Mississippi. No hardhats were required, just a straw hat to keep the sun off. Below, the two-track ferry is being loaded with freight cars. The cluster of smokestacks in the background is the Great Lakes Carbon factory near the MoPac's Ivory Street Yards.

The *Willard V. King* was originally owned by the New Orleans, Texas and Mexico and named the *B. F. Yoakum* after the Frisco railroad's president who owned the Gulf Coast Lines at the time. It was built in 1910 at Dubuque, Iowa, by the Dubuque Boat and Boiler Works. In 1913, the ferry was renamed *Willard V. King* after the new president of the re-organized railroad after Frisco lost control of it. Initially, the two-track ferry worked between Baton Rouge and Anchorage, Louisiana. In 1927, it was brought to St. Louis to replace the aging *James P. Lockwood* ferry to work the "Ivory Transfer" (Ivory Street incline to East Carondelet, Illinois) until January 1940 when the Municipal Bridge was opened. MPRR photographs

Missouri Pacific formed their bus line subsidiary "Missouri Pacific Transportation Company" in September 1928 and within six months the company had 1,800 route miles and 125 buses. In November 1935, the company joined the National Trailways Bus System with routes covering most of the railroad's territory. The 679 is shown here in Trailways red and cream colors with black trim at Little Rock in 1941.

The 679 was part of a group of eighteen Model 732 buses (Yellow's first rear-engine buses) the company bought from Yellow Coach being delivered in December 1937 and January 1938. In 1948, the MPTCo quit the Trailways system and partnered with Greyhound. Later, on January 1, 1957, the bus company was sold to several private bus companies. MPRR photograph

In addition to owning rail ferries and a bus line, the Missouri Pacific owned a busy electric railroad between Houston and Baytown, Texas, from 1927 to 1948. Car 524 was built by American Car Company in 1927 costing $16,418. Named "Highlands" when owned by the Houston North Shore Railway, the Missouri Pacific numbered it 524 after buying the railroad. It was double ended by the MoPac in 1931 to ease passengers entering and exiting the car. To illustrate how busy this line was, the July 3, 1943, employee timetable showed the HNS was running thirty-seven daily scheduled passenger trains and four local freights. Here 524 makes a station stop at the Highland Station in 1948. The electric car's last run on the HNS subsidiary was on September 25, 1948, being replaced by red and cream colored railbuses. Charles C. Robinson photograph

Wreck just east of Lake Junction on the mainline in 1946. The cause was an engineer on an eastbound freight exceeded the speed limit. Two steam wreckers are at work pulling the jumble of freight cars up the mainline right of way northside of the embankment. In the top view the photographer is looking east and the bottom image he's looking west. Wayne Leeman photographs

Between 6:35 pm on June 10 and 6:20 am on June 11, 1938, the area near Hematite, Missouri, received almost eight inches of rain, causing Joachin Creek to reach flood stages and overflow.

Water covered the tracks from a depth of three to five inches, washing out the tracks for five pole lengths. Southbound No. 7, "The Southerner," with locomotive 5344, four express cars, one baggage–mail car, four mail cars, two coaches, two Pullmans, and one dining car had left St. Louis at 2:02 am and ran into water at Hematite thirty-five miles farther south and derailed while traveling at fifteen to twenty miles per hour. The ballast and roadbed had washed out causing the accident.

The engine stopped on its right side, the tender was entirely submerged, and the first three cars were derailed to the right and were partially submerged. The waters were turbulent and Engineer Malloy was killed. Fireman McRill was able to save himself by perching in the top of a small tree where he was rescued 3½ hours later during which the water had risen 2½ to 3 feet. A section man reported to the dispatcher he'd covered the territory from Pevely to Riverside and while considerable water had fallen, the track was "all right." The last train, No. 56, had passed over this area between 12:45 and 12:50 am and No. 55 and No. 56 had met at Horine with neither train reporting any difficulty. Paychecks for the DeSoto shops were in one of the baggage cars and laid out at the offices so they could dry out and be distributed. Wayne Leeman photographs

Missouri Southern Brill Model 55 numbered M-100 is on train No. 1 ready to depart the MoPac Leeper depot for Bunker, Missouri, in December 1940. The Model 55 was a perfect fit for the fifty-four-mile mile trip to Bunker and the fifteen-mile branch line from Hobart to Brushy. It had a four-cylinder gasoline mechanical drive train engine that cost only sixteen cents per mile to operate. While the car was built to carry forty-three passengers and baggage, M-100 appears to have on this trip only a husband and wife with their canine companion. The MP and MS shared the depot at Leeper, Missouri, where the two roads also interchanged freight traffic. The Missouri Southern was abandoned on May 15, 1941. Wayne Leeman photograph

The two are placing the 1964 Chevrolet Bel Air on a hi-rail on the mainline at Poplar Bluff, Missouri. Just north of the station is the Railroad YMCA. In the photo below, Fletcher is getting ready to give a freight a "roll-by" inspection. The four railcars behind the locomotive are lead concentrate container cars. The lead concentrate was trucked to Middlebrook, Missouri, and loaded there (on specially designed railcars and containers for this service) destined to the lead smelter at Herculaneum. Wayne Leeman photograph

On October 18, 1965, Wayne Leeman took a hi-rail trip from Poplar Bluff to St. Louis with Missouri Division Superintendent Floyd Fletcher and Southern District Engineer Dean Barton. Fletcher, standing above to the right, began his career as an operator on the Eastern Division and later retired as a general manager in Chicago. Barton started in 1953 on the Missouri Pacific as an instrument man and was promoted to roadmaster, trainmaster, and district engineer. When D. B. Jenks came to the MoPac in 1961, officers were "encouraged" to wear dark suits and a hat.

Bibliography

Adams, Walter M. *The White River Railway Being a History of the White River Division of the Missouri Pacific Railroad Company 1901-1951*. The Ozark Mountaineer, 1991.

A.F.E Book 1939-1951. Reproduced by the Missouri Pacific Historical Society.

Collias, Joe G. *In Search of Steam*. First edition. Howell-North Books, January 1, 1972.

Collias, Joe G. *Mopac Power*. First edition. Howell-North Books, January 1, 1980.

Collias, Joe G. *Texas & Pacific Power*. MM Books, March 1, 1990.

Corporate History of the Missouri Pacific Railway Company and the St. Louis, Iron Mountain & Southern Railway Company in accordance with Valuation Order No. 20 Interstate Commerce Commission May 13, 1915.

Duckworth, Charles A. *The Missouri-Illinois Railroad Missouri Pacific's Route Through the Lead Belt and Little Egypt*. Missouri Pacific Historical Society, 2020.

EuDaly, Kevin. *Missouri Pacific Diesel Power*. First edition. White River Publications, February 1, 1998.

Grain Elevators of North America. Fifth edition. Grain & Feed Journals, Chicago, Illinois, 1942.

Leeman, Wayne. "Kirkwood Hill." *Trains* magazine (July 1943).

Patterson, W. J. *ICC Investigation 2275 Covering the Missouri Pacific Derailment at Hematite, Missouri June 11, 1938.*

Robinson, Charles and Paul DeVerter, *Houston North Shore, Bulletin 122 of the Central Electric Railfans Association* (2000).

Stagner, Lloyd. *Steam Finale in Kansas*. South Platte Press, January 1, 1992.

Stout, Greg. *Route of the Eagles 'Missouri Pacific in the Streamlined Era'.* First edition. White River Publications, February 1, 1998.

Thomas, Larry. *Towers in St. Louis, A Historical Guide to Interlocking Towers in the St. Louis Area, Terminal Railroad Association of St. Louis Historical and Technical Society Summer 2008 – Issue 68.*

Various issues of *The Eagle*, a quarterly magazine published by the Missouri Pacific Historical Society from 1974 to 2021.

Various Missouri Pacific employee and public timetables.

Index

About the Author

Charlie Duckworth graduated from Fort Hays University in 1973 and started with the Missouri Pacific Railroad (MoPac) in 1974, retiring from Union Pacific after thirty-eight years having held several management positions in the IT, Operating, and Marketing Departments. He and his wife Linda live in Omaha, Nebraska, and have been married forty-four years. They have three grown children (Anne, Emily, and Colin) and six grandchildren. His father worked for the "MoPac," so the family had a pass and frequently rode the Eagle passenger trains for free. These trips probably sowed the seeds for Charlie's interest in the Missouri Pacific and its history.

Charlie is a life member of the Missouri Pacific Historical Society and has served as its president on several occasions and on its board as a director for several years. He has had several published articles on railroad history over the years with this being his third book.